COUN
ASSISTANT:
—CONTAINING—

A Collection of NEW DESIGNS of

CARPENTRY AND *ARCHITECTURE*;

Which will be particularly useful, to Country Workmen in general.

ILLUSTRATED WITH *NEW* AND *USEFUL DESIGNS* OF

Frontispieces, Chimney Pieces, &c. Tuscan, Doric, Ionic, and Corinthian Orders, with their Bases, Capitals, and Entablatures: Architraves for Doors, Windows, and Chimneys: Cornices, Base, and Surbase Mouldings for Rooms: Doors, and Sashes, with their Mouldings: The construction of Stairs, with their Ramp and Twist Rails: Plan, Elevation, and one Section of a Meetinghouse, with the Pulpit at large: Plans and Elevations of Houses: The best Method of finding the length, and backing of Hip Rafters: Also, the tracing of Groins, Angle Brackets, Circular Soffits in Circular Walls, &c.

CORRECTLY ENGRAVED ON THIRTY COPPER PLATES;
WITH A PRINTED EXPLANATION TO EACH.

By ASHER BENJAMIN.

PRINTED AT GREENFIELD, *(MASSACHUSETTS)*
By THOMAS DICKMAN.

M,DCC,XCVII.

APPLEWOOD BOOKS
BEDFORD, MASSACHUSETTS

The Country Builder's Assistant was originally published in 1797.

ISBN: 1-55709-104-8

Thank you for purchasing an Applewood Book. Applewood reprints America's lively classics—books from the past that are still of interest to modern readers. For a free copy of our current catalog, please write to Applewood Books, 18 North Road, Bedford, MA 01730.

10 9 8 7 6 5

Library of Congress Cataloging-in-Publication Data
Benjamin, Asher, 1773-1845.
  The country builder's assistant: containing a collection of new designs for carpentry and architecture, which will be particularly useful to country workmen in general : illustrated with new and useful designs of frontispieces, chimney pieces, &c. . . . correctly engraved on thrift copper plates; with a printed explanation to each / by Asher Benjamin.
     p.   cm.
  Originally published: Greenfield, Mass. : Thomas Dickman, 1797.
   ISBN 1-55709-104-8
   1. Architecture—Early works to 1800.   2. Carpentry —Early works to 1800.   I. Title.
NA2520.B4   1992
720–dc20   92-42123
                                                           CIP

PLATE 1.

FIGURE A. Scotia; to draw the curve line of the Scotia, divide the height into three parts; place one foot of your compasses in z, and extend the other to 3, and draw the Arch line $3\ b$; then move your compasses to a, and extend to b, and draw the curve line b, d.

FIG. B. Aftrigal; divide its breadth into five equal parts; give one to each Fillet, and three to its Round.

FIG. C. Bead, which is $\frac{1}{4}$ of a circle.

FIG. D. Quirk Ogee; divide its breadth into five parts; give two of those parts to the Hollow and three to the Round.

FIG. E. Quirk, Ogee and Aftrigal; divide its breadth into three parts; give one to its Aftrigal, and two to its Ogee.

FIG. F. Cimarecta; divide the half Radius into six equal parts, and with five of those parts, make points of interfection, c, c, which will be the centres of the curves.

FIG. G. Quirk, Ovolo and Aftrigal.

FIG. H. Cavetto and Aftrigal.

FIG. I. Ogee.

FIG. K. Ogee and Bead, which are all plain to infpection.

FIG. L. Keyftone; divide the face of the Keyftone, into 12 parts, and difpofe them to the Mouldings, as figured.

FIG. M, is a Banifter for Baluftrades, or any other place required; the diftance between the Banifters may be one half, or the whole breadth of the Banifter.

To proportion Architraves to Doors, Windows, &c. divide the width of your Door or Window, into seven or eight parts, and give one to the width of the Architrave: Divide that into the same number of parts, as are contained in the Architrave you make use of, if a Frieze and Cornice to the Door, give the Frieze equal to the width of the Architrave; or it may be one fourth or one third wider, the Cornice four fifths or five sixths of the Architrave.

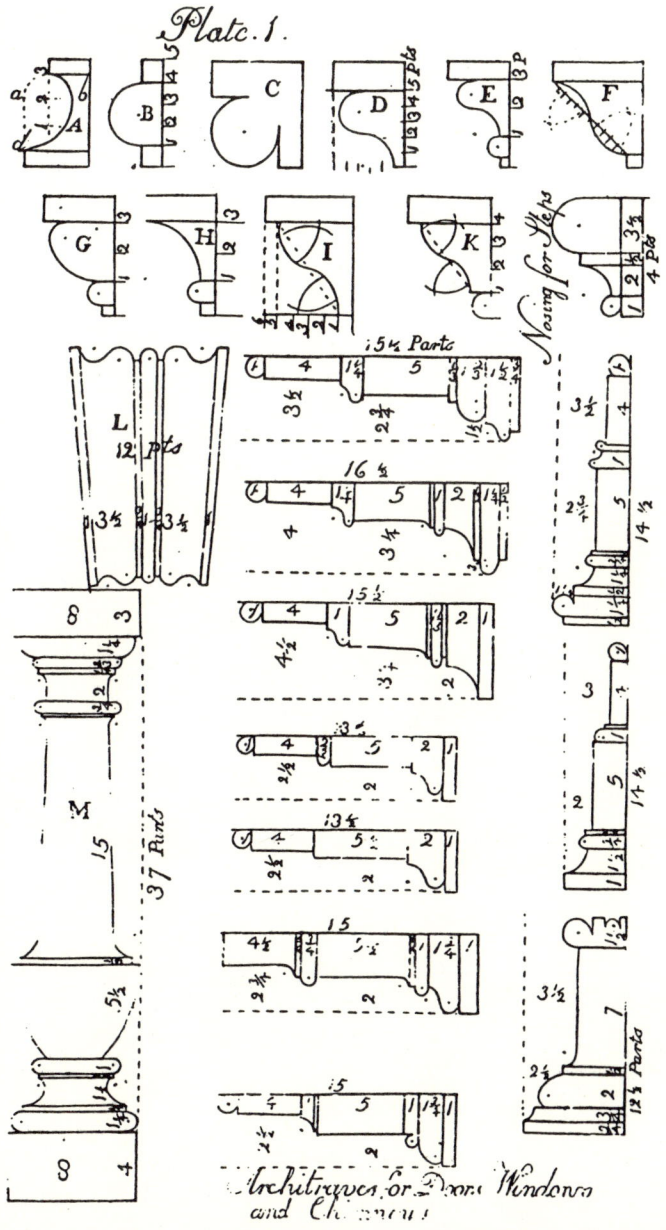

Plate 1.

Architraves for Doors, Windows and Chimneys

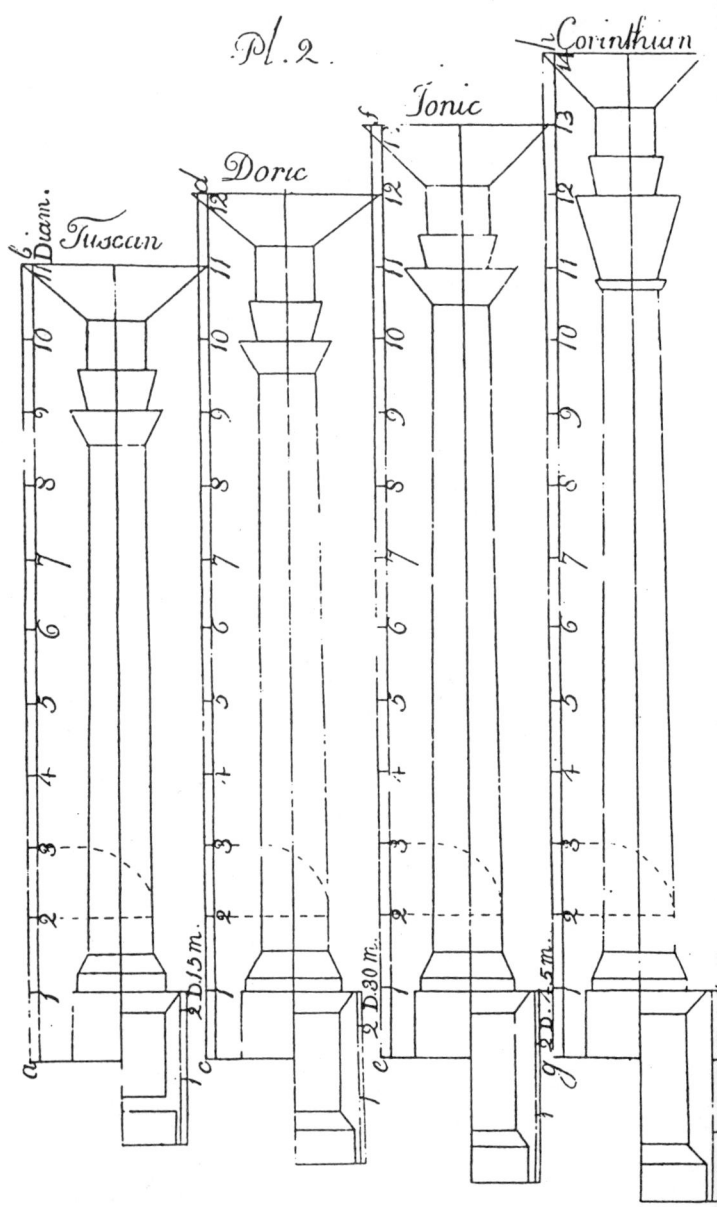

PLATE II.

To proportion the Tuscan Order, to any given height, on a Sub-plinth.

SUPPOSE the rod $a, b,$ to be the height given; divide it into 11 parts; one is the diameter of the Column at bottom; give one to the Sub-plinth, and two to the Entablature, or one diameter and 55 minutes. If a Pedestal is required, divide the given height into 49 parts, four of which will be the diameter of the Column, give nine to the Pedestal, and eight to the Entablature.

To proportion the Doric order, on a Sub-plinth, to any given height, divide the rod $c, d,$ into 12 equal parts; one is the diameter of the column at bottom; give one to the Sub-plinth, and two to the Entablature; the column, including Base and Capital, is nine diameters. If a Pedestal is required, divide the given height into 27 parts; two is the diameter of the column; give five to the Pedestal, and four to the Entablature.

To proportion the Ionic order to any given height, on a Sub-plinth, suppose the rod $e, f,$ to be the height given; divide it into 13 equal parts, one will be the diameter of the column, at bottom, two for the height of the Entablature; so that the column, including the Base and Capital, is 10 diameters high. If a Pedestal is required, divide the given height into 59 equal parts; four is the diameter of the column; give 11 to the Pedestal, and eight to the Entablature.

To proportion the Corinthian order to any given height, on a Sub-plinth, suppose the rod $g, h,$ to be the height given; divide it into 14 equal parts; one is the diameter of the column; give one to the Sub-plinth, and two to the Entablature; so that the column, including Base and Capital, is 11 diameters, Base 30 minutes, Capital 70 minutes. If a Pedestal is required, divide the given height into 16 parts; give three to the Pedestal, and two to the Entablature.

PLATE III.

The Tuscan Base, Capital, and Entablature, with all the Measure, figured from the Scale a, b.

THE Scale *a, b,* is one diameter of the column at bottom; divide the Scale *a, b,* into 12 parts, and one of them into five parts; then the Scale *a, b,* is divided into sixty parts, and those parts are to be disposed to the Mouldings, as figured, in height and projection, the projection set from a plumb line, dropped from the extreme of the Mouldings; *a* Sub-plinth one diameter; *b,* Base to column, 30 minutes; *c,* Capital to Column, 36 minutes; *d,* Architrave 34 minutes; *e,* Frieze 39 or 44 minutes; *f,* Cornice 42 minutes; the Scale *g,* to find the diameter of the Tuscan Column, to any given height, on its own Sub-plinth, the line *h, f,* is divided into 10 feet, and one foot into 12 parts, which are inches. Admit the given height to be eight feet, open your compasses from *n* to *o,* against eight feet on the Scale; then apply your compasses to the Scale of inches, from three to four feet on the Scale, you will find the diameter to be about nine inches and $\frac{5}{8}$; look for the height on the line *b, f,* and against the height, open your compasses from the line *n,* to the line *o*; then apply the compasses to the Scale of inches, which is the diameter sought for. If it is required to find the diameter of the column alone, open your compasses from the line *n,* to the dotted line *p.*

NOTE. These directions will answer for the scales laid down in the Doric, Ionic and Corinthian orders.

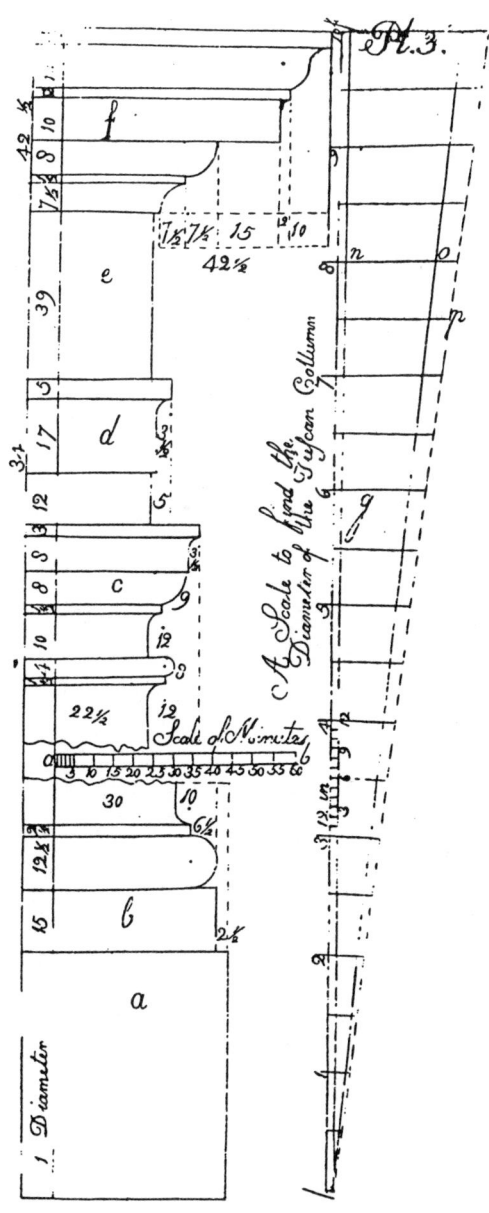

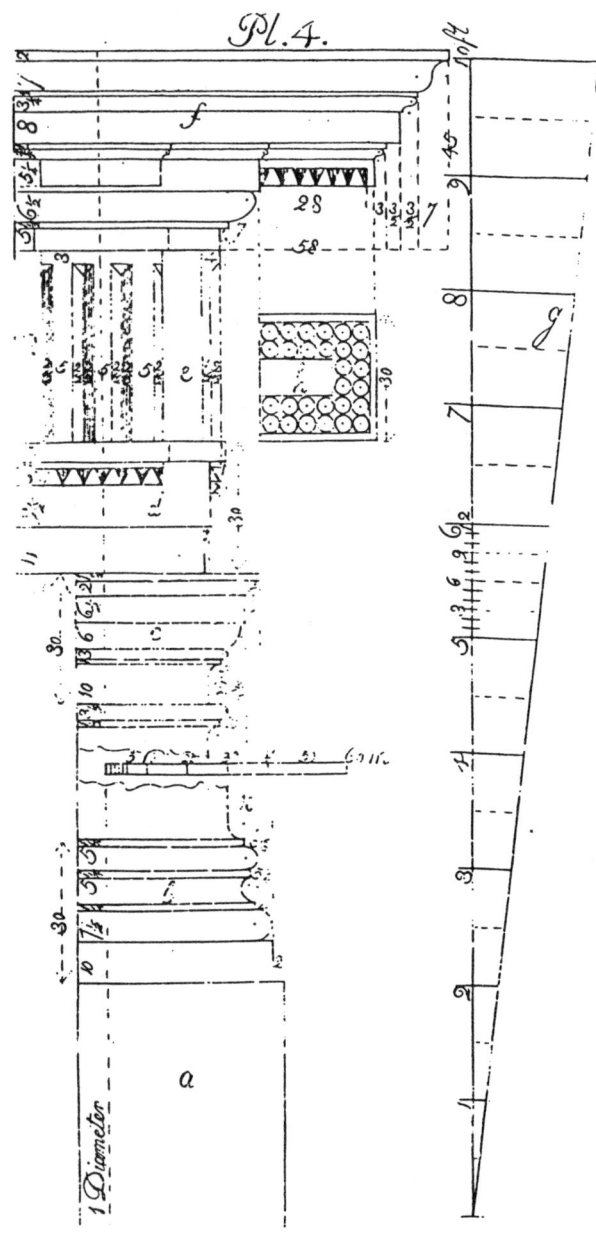

PLATE IV.

The Doric Base, Capital, and Entablature, with all the Mouldings, figured for Practice.

IN height and projection, the shaft of the column diminishes one sixth, that is, 60 minutes at bottom and 50 at top ; *a*, Sub-plinth one diameter; *b*, Base to column, 30 minutes ; *c*, Capital, 30 minutes ; *d*, Architrave, 30 minutes ; *e*, Frieze, 45 minutes ; *f*, Cornice, 45 minutes in height. The width of the Triglyph in Frieze, 30 minutes ; the distance from centre to centre, 75 minutes ; the interval between the Triglyphs, 45 minutes ; the width of the Triglyphs, 30 minutes, is to be divided into 12 parts, each part $2\frac{1}{2}$ minutes, that is, $2\frac{1}{2}$ minutes to each semi gutter, and 5 minutes to each fillet, as figured, the profile or thickness of the Triglyph, is $3\frac{1}{2}$ minutes, $2\frac{1}{2}$ minutes to the depth of each gutter and one to the bottom.

IN Intercolumniations, for Porticoes, Colonades, Arcades, &c. due regard must be had to the number of Triglyphs and Mutules between the central line of Columns in the Doric Order, 3 diameters 45 minutes, take 3 Triglyphs ; 5 diameters take 4 Triglyphs ; 6 diameters 15 minutes, take 5 Triglyphs ; 7 diameters 30 minutes, take 6 Triglyphs ; 8 diameters 45 minutes, take 7 Triglyphs, &c.

PLATE V.

The Ionic Base, Capital, and Entablature, with all their Mouldings, figured for practice, in height and projection.

THE height of the Base 30 minutes; the height of the Capital 30 minutes, the height of the Architrave, Frieze and Cornice, 2 diameters; Architrave 35 minutes, Frieze 40 minutes, Cornice 45 minutes. A profile of modillion, to draw the curve or plancere of the modillion, divide the plancere of the modillion into 6 equal parts; then place one foot of your compasses at 2, on the dotted line, extend the other foot to 1, on the same line, then draw the quarter of a circle, as 1, 1; then set down one part and a half, as at *a*, and place one foot of your compasses at *a*, and extend the other to 1, and draw the arch line 1, to the line *a b*; then set your compasses at *b*, and draw the arch line to 5, which compleats the curve of the modillion. From centre to centre of the modillions 31 minutes, the width in front 10 minutes, the interval 21 minutes. Intercolumniations in the Ionic Order, 4 diameters 8 minutes from centre to centre of the columns, take 8 modillions; 5 diameters 10 minutes, take 10 modillions; 6 diameters 12 minutes, take 12 modillions. If dentils are used instead of modillions, the space from centre to centre of the dentils, is $7\frac{1}{2}$ minutes, and the interval between 2 minutes and a half, the dentil in front 5 minutes.

B, Base of Column; C, Capital; D, Architrave; E, Frieze; F, Cornice.

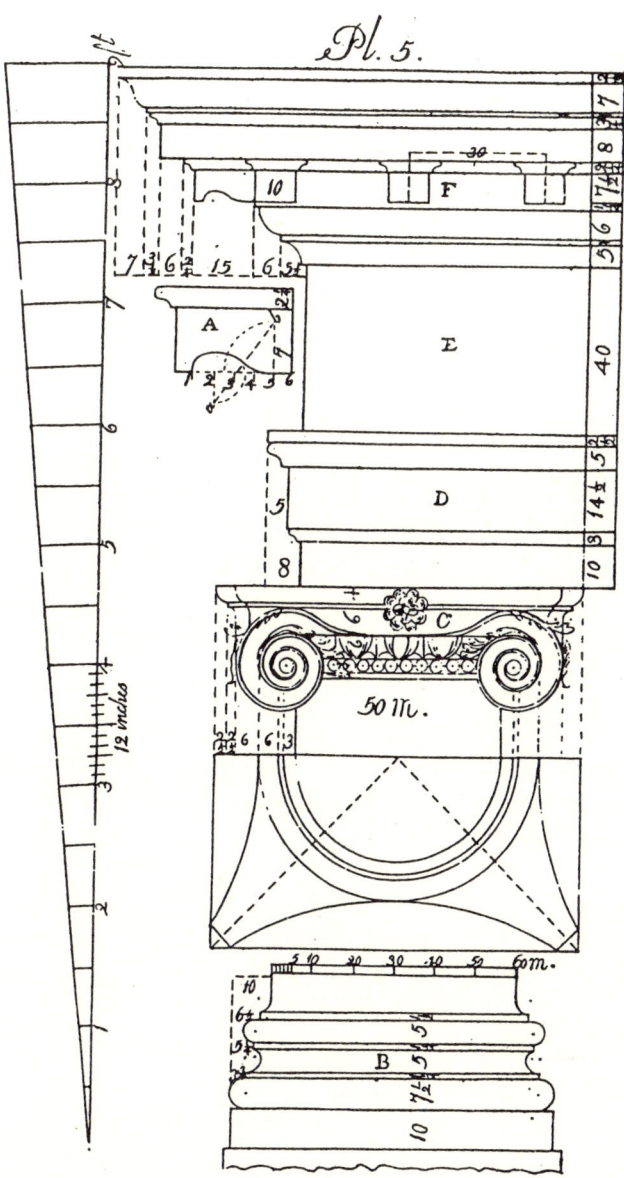

Pl. 5.

Pl. 6.
Part of the Ionic Cap.

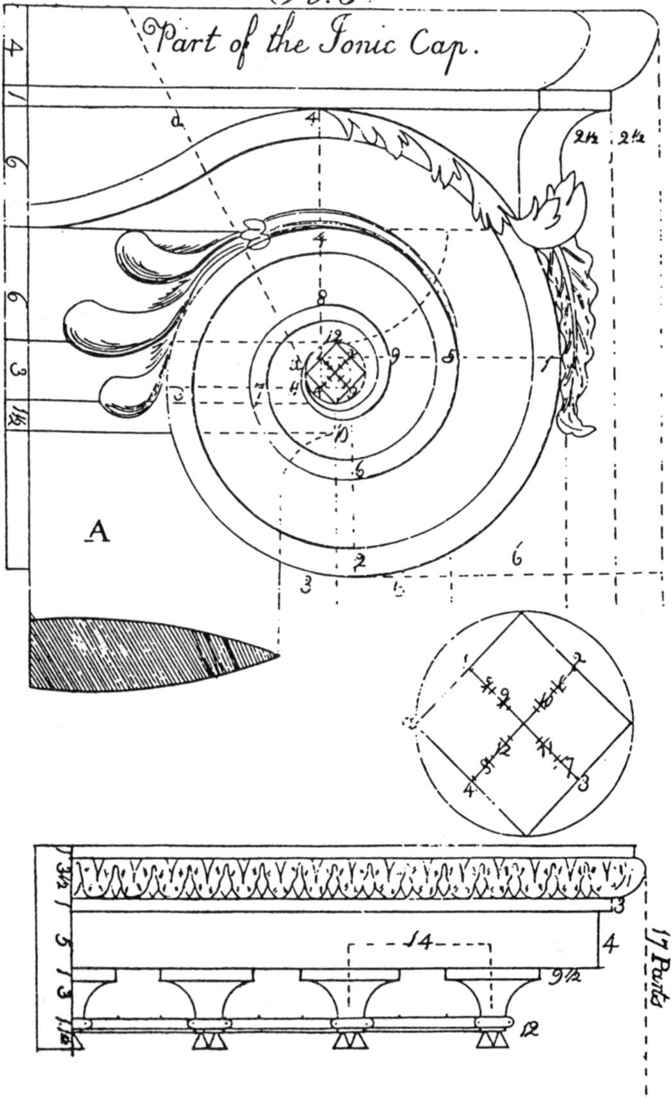

PLATE VI.

The Ionic Volute, with all the measures, figured for practice.

TO draw it, set the compasses at the angle *a*, in the profile, and draw the arch from 4 to *a*, on the back of the list, then draw it from *a*, down to the Ovolo on the centre. To draw the other part of the Volute, set one foot of the compasses at 1, on the side of the square of the eye, and extend to 4 under the fillit of the Abacus, and turn round to 1, opposite 1, on the side of the square where your compasses were first set ; then set your compasses on the other side of the square at 2, and draw the arch 1, 2 ; then set the compasses on the other side, at 3, and draw the arch from 2 to 3 ; then set the compasses at 4, and draw the arch line 3, 4, which is one revolution ; then take the centre 5, and draw the arch line 4, 5 ; then take the centre 6, and draw the arch line 5, 6 ; then the centre 7, and draw the arch 6, 7 ; then the centre 8, and draw the arch 7, 8 ; next the centre 9, and draw the arch 8, 9 ; and so on, for the rest you see in the eye at large, Fig. *a*, the small lines within the first centres, these are the centres for the inside of the list, to give its diminishing.

PLATE VII.

The Corinthian Base, Capital, and Entablature, with all the Mouldings figured for practice.

A, BASE Mouldings to Column; B, Architrave; C, Frieze; D, Cornice. The Corinthian Modillion $11\frac{1}{4}$ minutes in front, and 35 from centre to centre; Base Mouldings $32\frac{1}{2}$ minutes; height of Capital 70 minutes; Architrave 35, Frieze 37, Cornice 48 minutes; 4 diameters 40 minutes; from centre to centre of the columns, take 8 Modillions; 6 diameters 25 minutes, take 11 Modillions; 7 diameters take 12 Modillions.

Pl. 7.

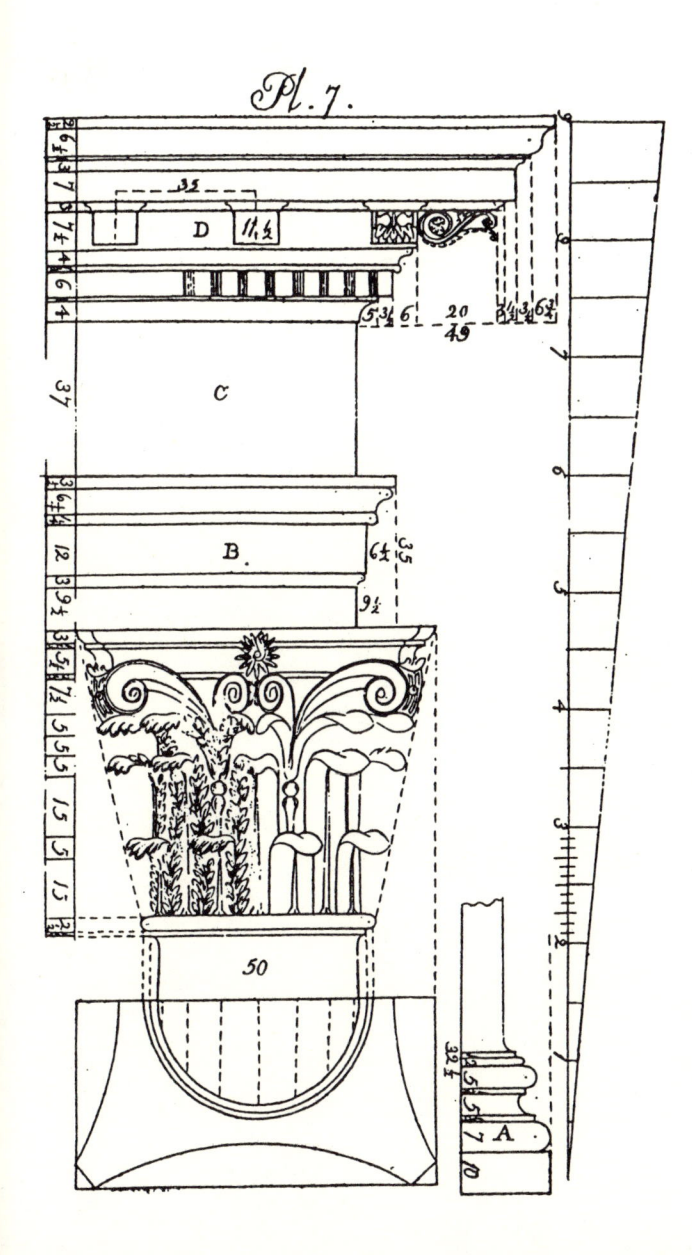

Pl. 8.

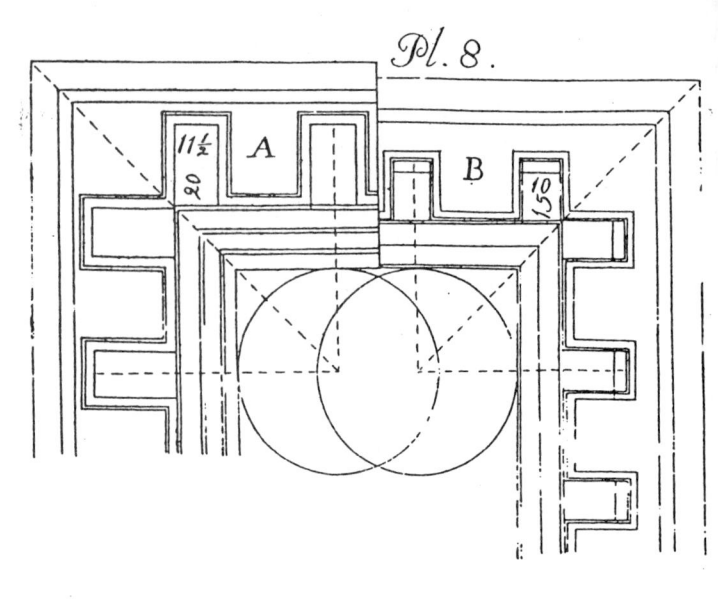

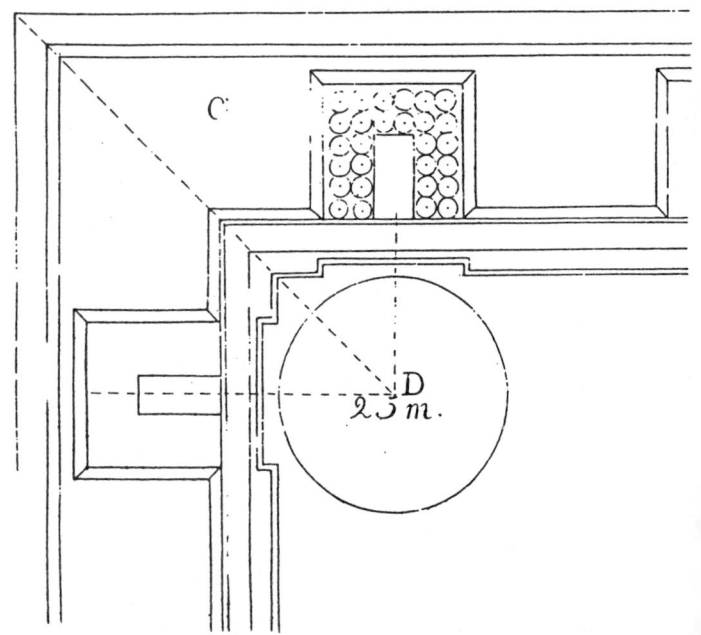

11½ A
20

B 10
15

C

D
25 m.

PLATE VIII.

FIGURE A, Plancere of the Corinthian Cornice.

Fig. B, Plancere of the Ionic Cornice.

Fig. C, Plancere of the Doric Cornice.

Fig. D, Neck of Column.

PLATE IX.

FIGURE 3 shows the diminishing of the shaft of a column; divide the length of the shaft into 3 equal parts; then divide the two upper third parts into 3 parts, 1, 2, *a*; then make the half circle at *b*, and divide the line 3, *b*, into 12 parts, each part equal to 5 minutes, then take 10 parts in your compasses, and prick down at *a*, it being the top of the column, and likewise on the circle at *c*; then divide the curve line into 3 parts, as 1, 2, 3; then take the line 1, in your compasses, and prick on the column at 2; then take the line 2 in your compasses and prick on the column at 1; then prick in nails at *b*, 1, 2, *a*; then hold a thin slip true to the first third part of the column, *b*, *d*; then bend it round to *a*, and mark as that curve directs, which will be the diminishing of the column.

Fig. 1, Shows how to set out the Flutes and Fillits on a column; take the Girt at bottom, and extend from *a* to *b*, on Fig. 1, likewise the Girt at the neck, and extend from *c* to *d*, and mark the Flutes and Fillits, as from *a* to *b*, on a slip of strong paper; fix it tight round the column, and mark them on the column, run 96 parts on a right line, as *e*, *f*, which must be less than the circumference of the column. To set out the Flutes and Fillits, on a Pelaster, run 29 parts on the line 1, 2.

Fig. 2, Greater than the diameter make the triangle 1, 2, 3, by setting the compasses at 1, 2, and turning them to 3, draw lines to 3, then the Pelaster, *a*, *b*, is divided; for the Flutes and Fillits, give 3 to a Flute and 1 to a Fillit.

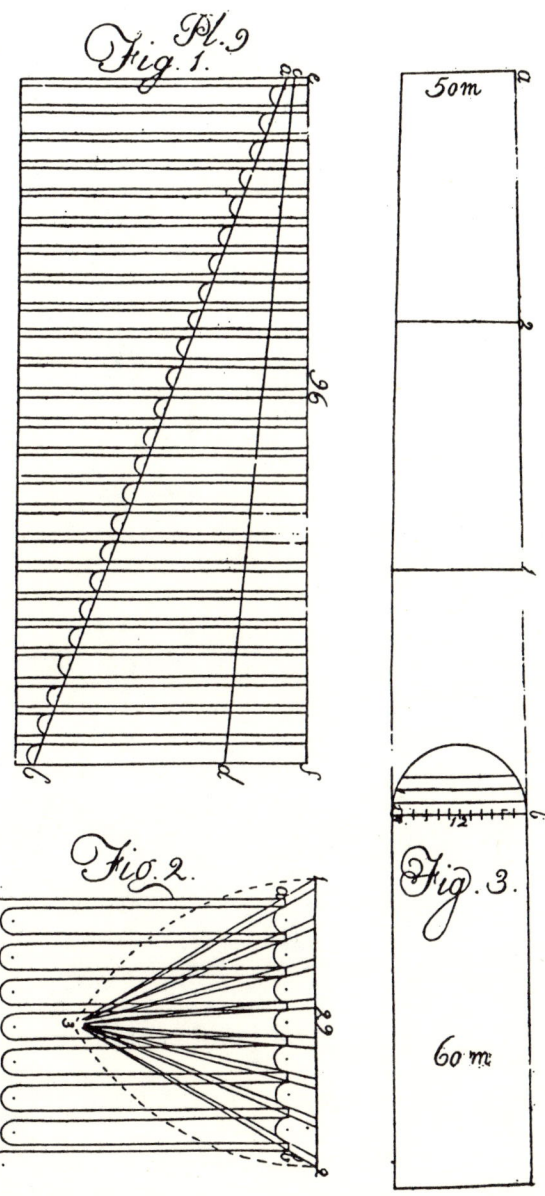

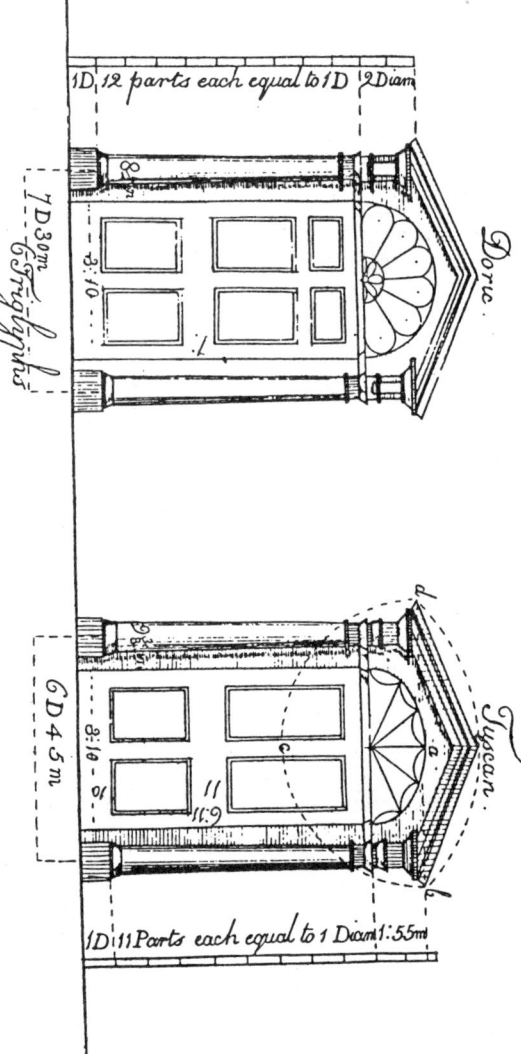

Pl. 10.

PLATE X.

TUSCAN Front, drawn one quarter of an inch to a foot, the clear passage 3 feet 10 inches, the height 6 feet 11 inches, the height of column 7 feet 1 inch, to be divided into 9 equal parts, one of which parts will be the diameter of the column at bottom; give one of them parts to the Sub-plinth; the distance from centre to centre of the column is 6 diameters and 45 minutes. To find the pitch of the Pediment, set the compasses at *a*, in the Tympan of the Pediment *d, c, b,* and draw the arch line *b, c, d;* then set the compasses at *c,* and draw the arch line *d, e, b,* which gives the height of the Pediment at *e.* This method gives the pitch to any Pediment.

DORIC Front, the clear passage 3 feet 10 inches, door 7 feet high; divide the height of the column into 10 parts, one of which is the diameter of the column; give one diameter to the Sub-plinth, and two to the Entablature; the distance from centre to centre of the columns, is 7 diameters and 30 minutes, which will take 6 Triglyphs.

PLATE XI.

IONIC and Corinthian Fronts, drawn one fourth of an inch to a foot, wit all their parts figured for practice, which is plain to inspection.

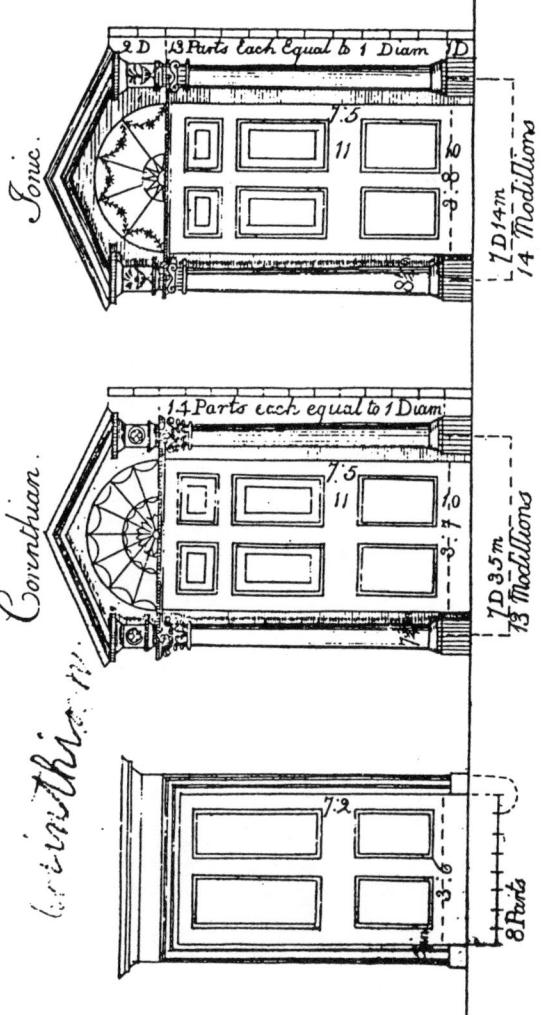

Pl. 11.

Pl. 12.

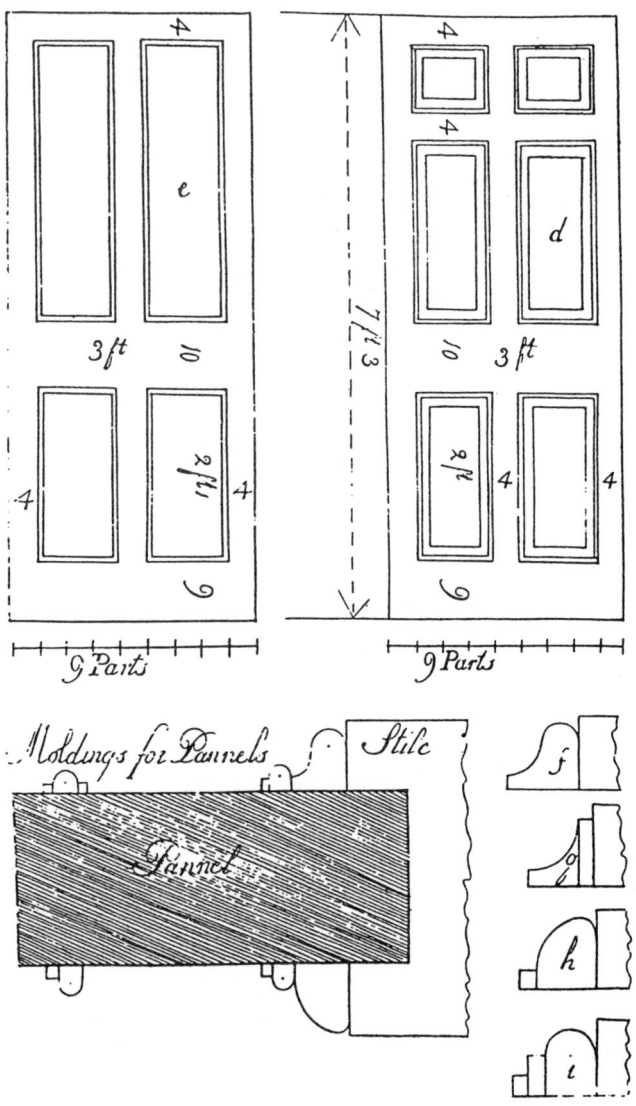

PLATE XII.

ON this plate is a four and six pannel Door, with all the meafures exactly figured. Divide the width of the Door into 9 parts, one of which is the width of the margin of the ftile; *i, h, g, f,* are mouldings for Doors, full fize for practice.

PLATE XIII.

FIGURE A, is an end view of a bar for an Ovolo Sash, full size for practice.

Fig. B, Do. Astrigal.

Fig. C, Do. Astrigal and Hollow.

Fig. D, An end view of a Stile for Fig. C, full size for practice.

Fig. E, Shows the profile of the Stiles to the top and bottom Sash. The shaded part shows the Dove tail for the Meeting Rails, which is plain to inspection.

Pl. 13.

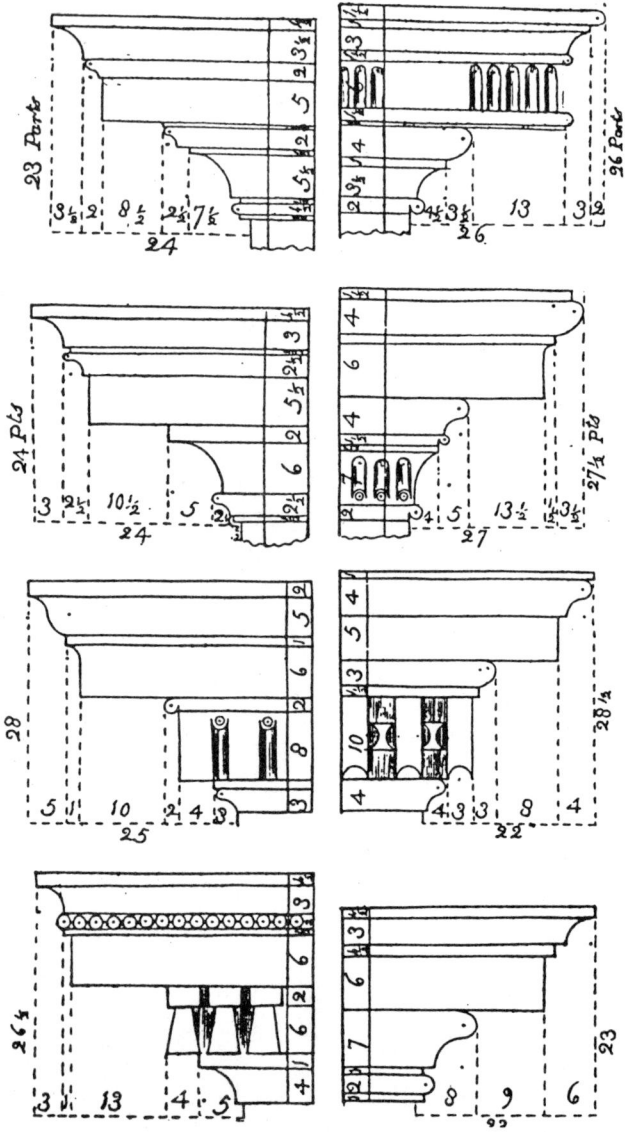

Pl.14.

PLATE XIV.

To proportion Cornices to Rooms or any other place required,

DIVIDE the whole height of the Room into 22, 24, or 26 parts, and give one of those parts to the Cornice, which is to be divided into the same number of parts as are contained in the Cornice you make use of, and those to be disposed to the Mouldings, in height and projection, as figured on the Plate. If those Cornices are used on the out side of buildings, divide the height into 19 or 20 parts, one of which will be the height of the Cornice. If the Cornices of any of the Orders are used without the Architrave and Frieze, the height may be one nineteenth, or one twentieth, as above, and divided into as many parts as there are minutes contained in the whole Cornice, which are to be disposed according to the directions in each Order.

PLATE XV.

On this Plate is the Tuscan, Doric, and Ionic Pedestal Mouldings, which may be drawn from the same scale, that you draw the orders from.

TO proportion Base and Surbase Mouldings, to the Pedestal parts of Rooms, the height from two feet six inches, to two feet ten inches to top of the capping, divide the height into nine or ten parts; give one to the Surbase, one half or two thirds to the Base Mouldings, and one and one third to the Plinth: Or, Suppose the Surbase to be two feet eight inches from the floor, the Surbase will be three inches and one half, the Base one inch and three quarters, the Plinth $4\frac{1}{2}$ or 5 inches. Divide the one ninth, or the $3\frac{1}{2}$ inches into as many parts as are contained in the Surbase you make use of, and dispose those parts to the Mouldings, in height and projection, as figured on the Plate; and likewise the one half, or one inch and three quarters, to be divided into as many parts as the Base Moulding you make use of, and those parts to be disposed to the Mouldings, in height and projection, as figured on the Plate.

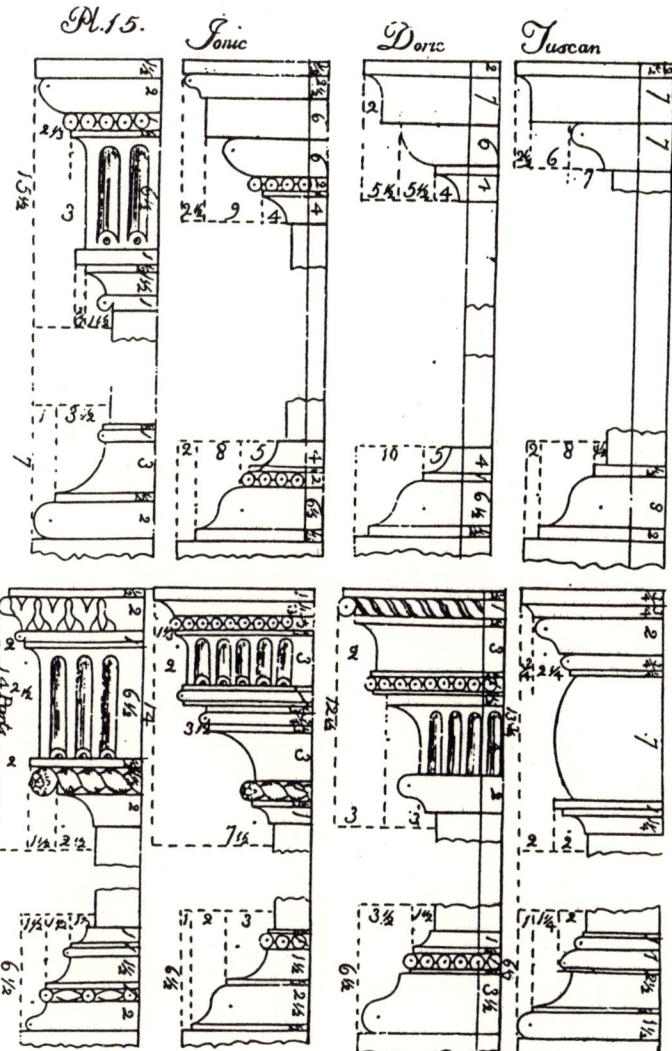

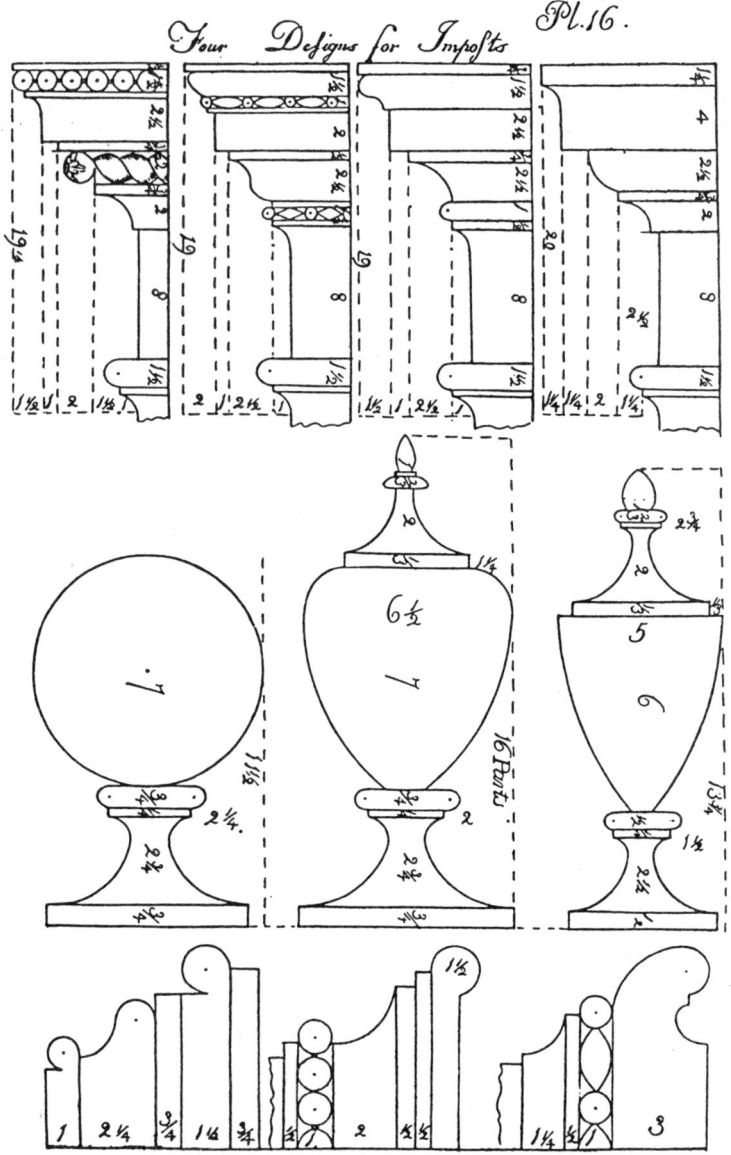
Four Designs for Imposts
Pl.16.

PLATE XVI.

To proportion Impost Mouldings to Arches.

FOR the height of the Impost, including the Necking, divide from the Floor to the Springing of the Arch, into 19 or 20 parts; take one for the Impost, including the Necking, and divide the height into as many parts as are contained in the Impost you make use of, and difpose those parts to the Mouldings as figured, in height and projection.

PLATE XVII.

FROM Plate 17 to Plate 20, are defigns for Chimney Pieces, drawn one half an inch to a foot.——Plate 17, is a plain Chimney Piece, with its Cornice, Architrave, Bafe, and Surbafe, drawn half fize for practice; divide the width, or opening of the Chimney, into eight or nine parts; give one eighth, or one ninth to the breadth of the Architrave.

Pl. 17.

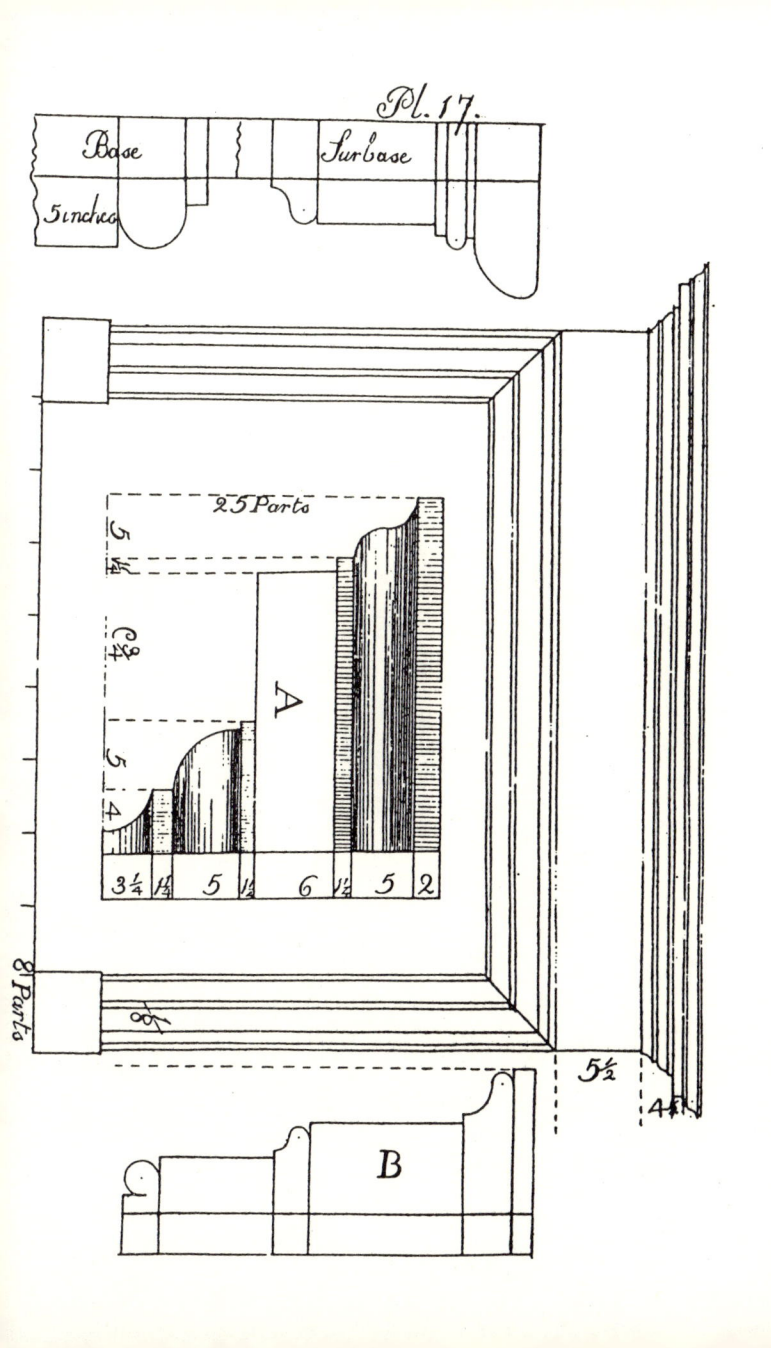

Base   Surbase
5 inches

25 Parts
5¼
6¾
5
4
3¼ 1¼ 5 ½ 6 ½ 5 2

A

8 Parts

5½ 4¼

B

Pl. 18.

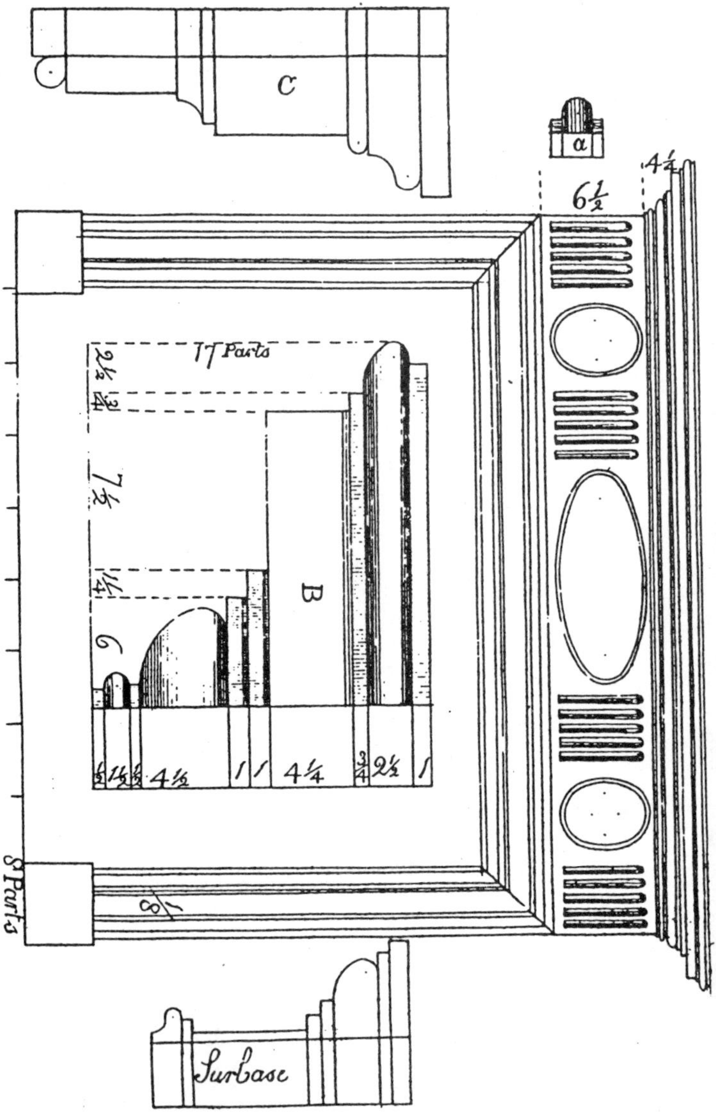

Surbase

PLATE XVIII.

B, CORNICE, half size; C, Architrave half size; A, Moulding round Tablet, full size for practice.

PLATE XIX.

A, CORNICE, half fize, for practice.

B, Band to Architrave, half fize.

C, Necking to Pelafter, half fize.

D, Bafe to Pelafter, half fize.

E, Surbafe Moulding, half fize.

Pl. 19.

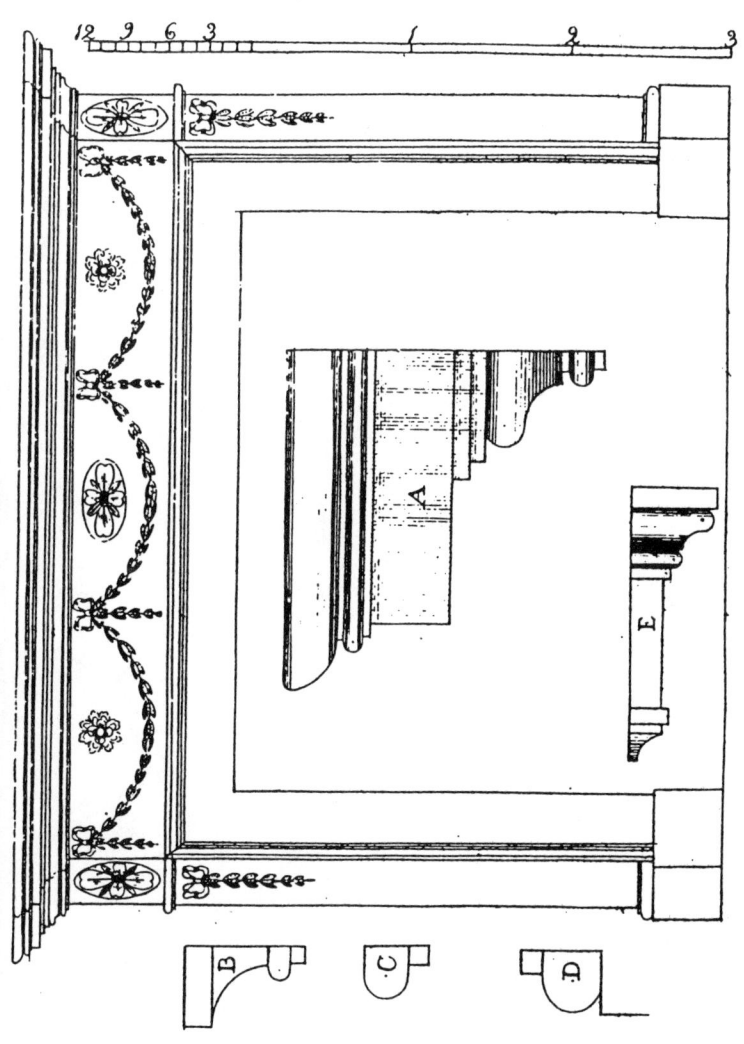

Pl. 20.

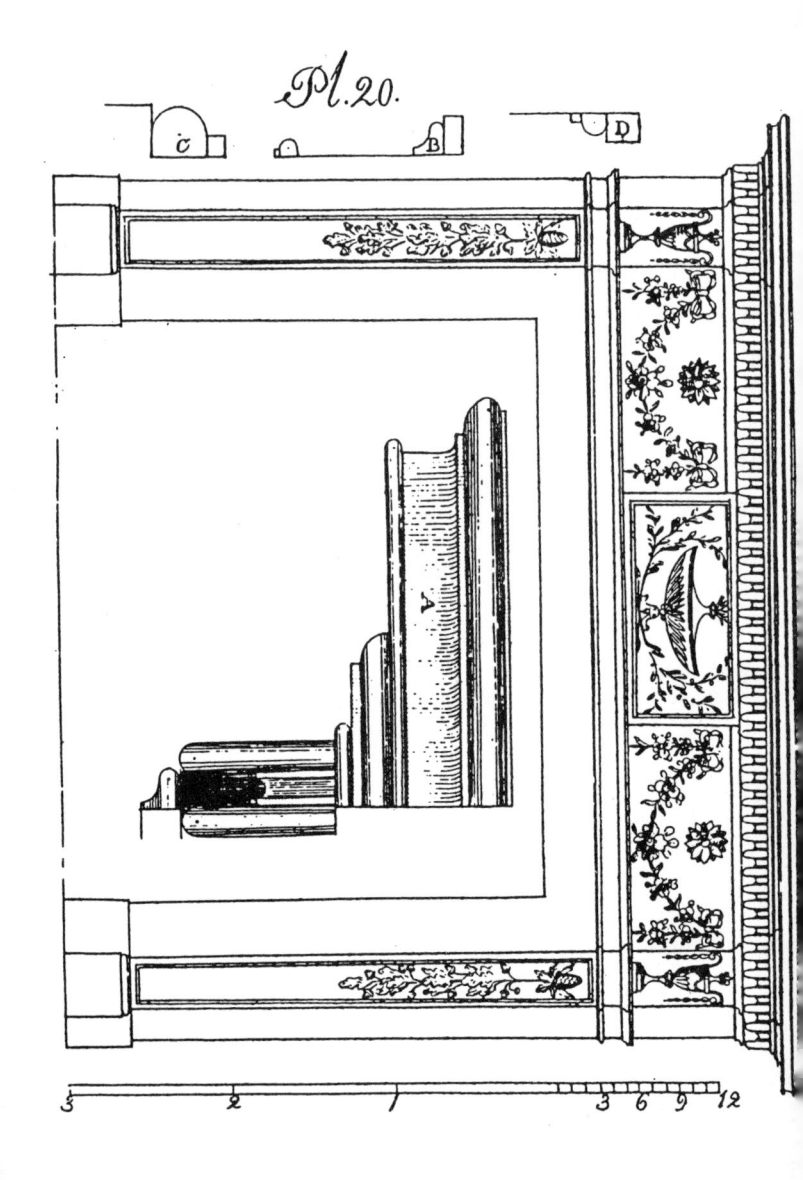

PLATE XX.

A, CORNICE, half size.

B, Necking to Pelaster, do.

C, Base to Pelaster, do.

D, The Moulding and Sinking of Pelaster, do.

PLATE XXI.

FIGURE 1, is the Plan and Scrole of the Twist Rail for a Staircase, to draw which, proceed in this manner: First, Describe a circle, whose diameter shall be equal to the breadth of two steps, *a*, *b*; then divide the circle into 8 parts and draw lines from the centre to each division; then draw the Chord line, *c*, *b*, which will be a fourth part of the circle; then take 2½ inches in your compasses, and describe a circle round the centre of the eye, whose diameter will be 5 inches; in the next place set one foot of the compasses on *c*, and extend the other to touch the outside of the circle last drawn, at *i*, and draw the arch line, *i*, *k*, which arch is to be divided into 8 parts, and a line drawn from the centre *c*, through each of those divisions, and continued to the line *i*, *b*, will be the diminishing scale.

To diminish the Scrole set one foot of the compasses on the centre of the eye, and extending the other to 1, on the scale *i*, *b*, turn it over to 1, on the first eighth part of the circle from *c*, next semi-diameter, from the centre of the eye, to 2 on the scale, and turn it over to 2, on the second eighth part of the circle; then the distance from the centre, to 3 on the scale, and turn it over to 3, on the third eighth part; then the distance from the centre, to 4 on the scale, and turn it round to 4, on the fourth eighth part of the circle, and this method is to be observed with respect to all the other divisions, until you come to the sixth, which will fall on the scale at six.

To find the centres for drawing each eighth part, take the distance from the centre of the eye, to *b*, at the top of the scale, and with that Radius, setting one foot of your compasses at *a*, on the side of the Rail where the twist begins, and describe a small arch, and with the same Radius, set the foot at 1, on the side of the Rail, and intersect the former arch, at *i*, which is the centre for the first eighth part, from *a* to 1; then set one foot of the compasses in the centre of the eye, and extend the other to 1, on the scale, and with that distance, setting one foot of your compasses at 1, on the edge of the Rail, describe a small arch, with the same Radius, placing the foot of your compasses at 2, on the edge of the Rail, intersect the former stroke on *a*

Fig. 1.

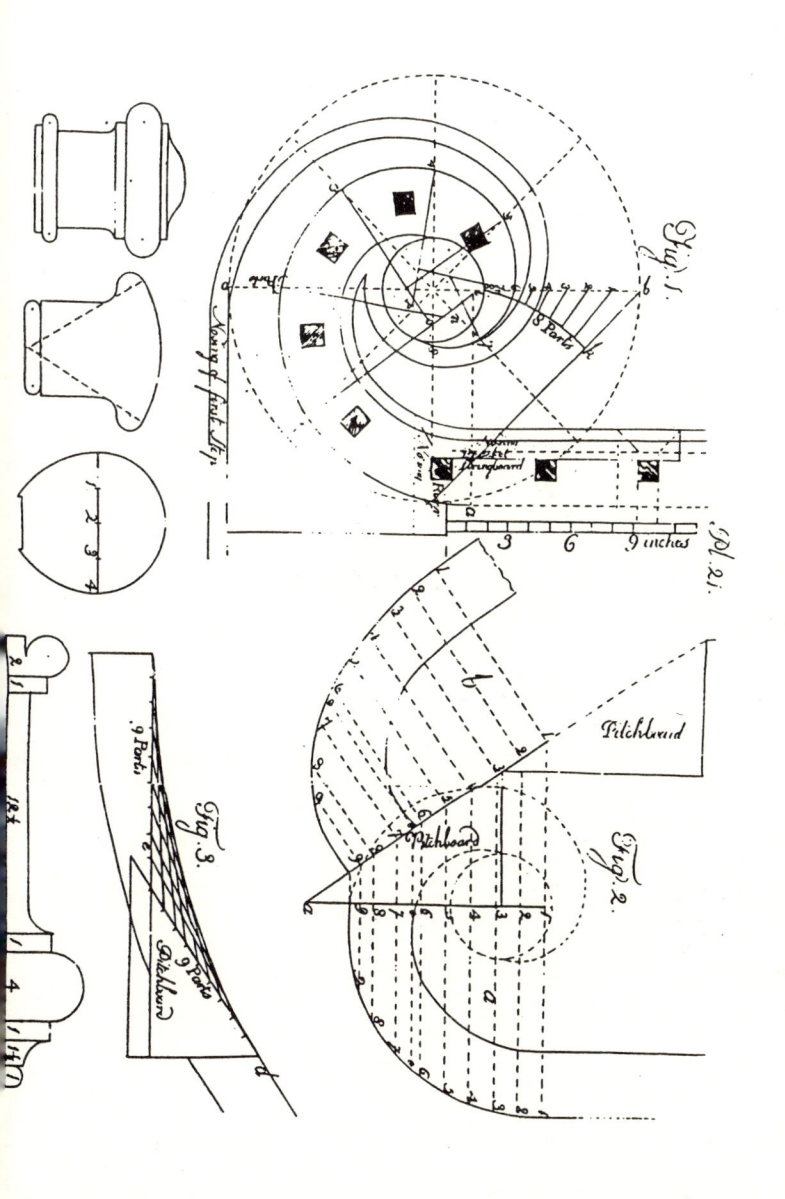

Fig. 2.

Fig. 3.

Pitchboard

Pitchboard

Pitchboard

9 Parts

9 Parts

8 Parts

Going of foot step

Pl. 2.

3 6 9 inches

PLATE XXII.

THE Elevation of a single Flight of Stairs, with all its parts figured, drawn by the scale below, of one inch to a foot, by which all the parts may be exactly measured. The next thing is to find the height of Hand Rails and Newels, with their Ramps and Knees; draw a line to touch the Nose of the Steps; then draw two other lines, at right angles with that, as *k, l, n, m*; then set on 2 feet 1 inch on each line as figured, which gives the height to the top of the Hand Rail; then set on the depth the Rail, *l, p, m, o*, and draw the line to meet the side of the Newel, at *e*, and to meet the side of the first Newel at *n*; next divide the steps for the Banisters, and find the place for the Banister on the first step, and draw a central line *q*, to meet the under side of the Rail at *t*; from the point of that meeting, draw the line *r*, or the under part of the Knee, then setting the depth of the Rail, as *r, s*, and drawing the line *s*, to meet the raking part of the Rail, at *t*, draw the line *t*; it will give the Mitre of the Knee, and the line *r* gives the height of the Newel. To find the centre for drawing the Ramp, set one foot of the compasses at the angle *e*; extend the other foot to *f*, and draw the arch line *f, g*, where the arch cuts the upper part of the Rail as at *h*, there is the place to hold the square, which will cut the level line, at *i*, the centre for drawing the Ramp. The length of the Newels are all equal, as figured, which may be proved by the scale. By this method, when the measures of any place is taken, any Staircase may be drawn by this 1 inch to a foot, by which will be found the length of String Boards, Newels, Banisters and Hand Rails, near enough for practice, in any case required.

which is the centre for the second eighth part; with the distance from the centre of the eye, to 2 on the scale, set the foot of your compasses at 2, on the edge of the Rail, describe a small arch, and removing your compasses to 3, on the edge of the Rail, intersect this arch at *o*, which will be the centre for the third eighth part; then take the distance from the centre to 3, on the scale, and setting the foot of your compasses at 3, on the edge of the Rail, describe an arch at *p*, which intersect from the point 4, will be the centre for the fourth eighth part, and so on, for all the rest. The outside of the rail may be drawn from the same centre of the inside.

To make the Raking Mould, for the turn of the twisted part of the Hand Rail, Fig. 2, first draw the plan of the Rail, Fig. 1, lay down the Pitch Board, parallel with the plan of the Rail, Fig. 1; divide the line *a*, 1, in the centre of the Scrole into 9 parts, and draw the dotted lines across the plan of the Rail, to the Raking line of the Pitch Board; then let those lines be drawn square from the Raking line of the Pitch Board; in the next place, from the plan *a*, take the distance 1, 1, and set it on the Raking Mould *b*; then take from *a*, 2, 2, and set it on the Raking Mould *b*; then take from *a*, 3, 3, and set it on the Raking Mould *b*; then take from *a*, 4, 4, and set it on the Raking Mould *b*; and so on for all the rest. The inside of the Raking Mould must be taken from the line *a*, 1, to the inside of the Rail, on the plan *a*.

FIG. 3, Is a Mould for the falling of the twisted part of the Hand Rail, from *a*, where the twist begins, to 3, on the edge of the Rail, where it ends, the remaining part of the Scrole being level. To make the Falling Mould, take the Base line of your Pitch Board, from the centre of the eye, to 2, on the edge of the rail, on the line *a*, *b*, Fig. 1; let the Raking line to the Pitch Board run at the same Angles, as the Pitch Board in Fig. 2; then trace from *a*, round on the edge of the Rail to 3, the twisted part of the Rail, and run those parts on the Raking line of the Pitch Board, until you get to the level line *e*; then run on the level line until you have got the length of the twist part of the Rail, *a*, 3, then divide the Raking line of the Pitch Board into 9 parts, and the level line into 9 parts, and by drawing right lines between those divisions, you will have the curve to the falling of the twist.

Pl. 22.

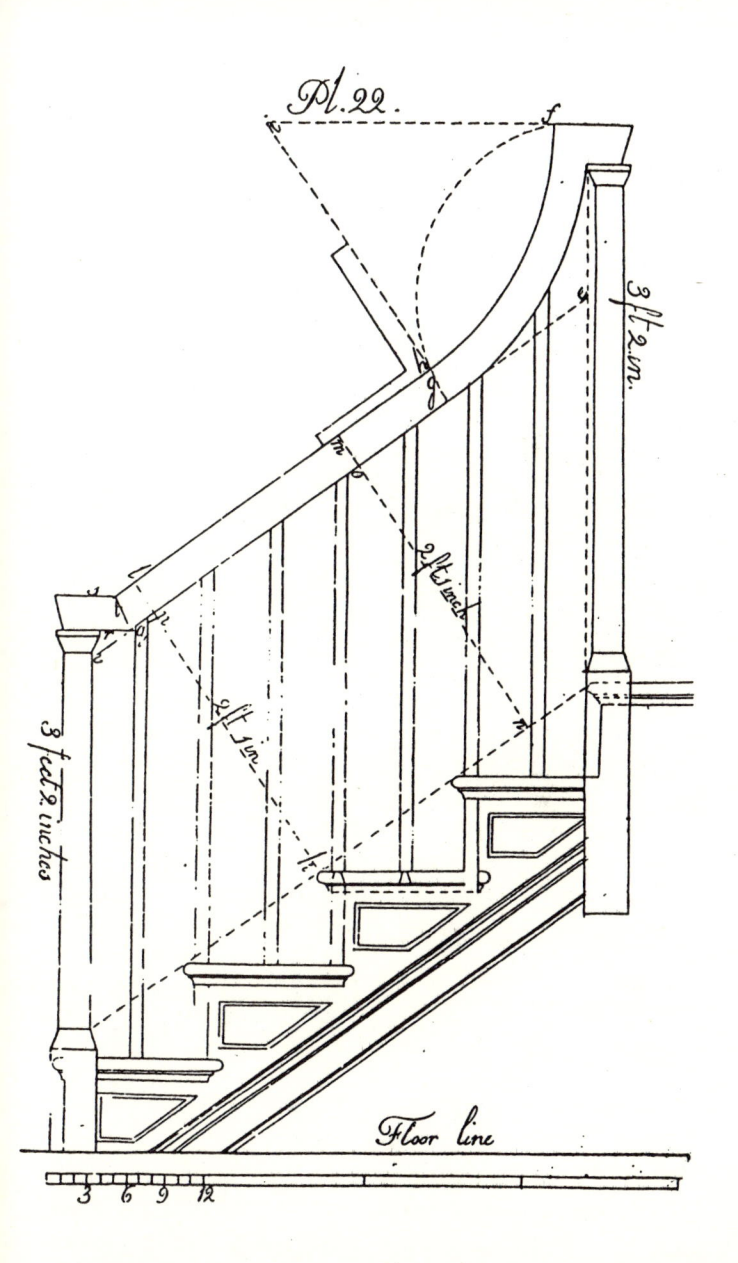

Pl. 23.

Scale of 2 Feet

PLATE XXIII.

A Plan of three Flights of Stairs, drawn from a scale of one inch to a foot.

THIS Plan is to show how to place the Newels and Banisters; the dotted lines are the Nosing of the Steps, and the other lines the faces of the Risers.

PLATE XXIV.

Of Raking Cornice for Pediments.

FIGURE A, Is a given Cornice, which the raking Cornice is traced from. Divide the level or given Cornice into 8 parts, and transfer them to the Raking Cornice B, as 1, 2, from the Cornice *a* to 1, 2, to the Cornice B and C, and so on, 3, 4, from *a*, to 3, 4, on *b* and *c*, and so on, 5, 6, to 5, 6; 7, 8, to 7, 8; 9, 10, to 9, 10; 11, 12, to 11, 12, the projections to be all alike.

Fig. 2, Is the end of a Raking Modillion, in a Pediment that contains three different Moulds; that of *a*, is the given Mould, which the other two are traced from, as 1, 2, to 1, 2; 3, 4, to 3, 4, projecting all alike.

Fig. 3. *a, b, d, c,* is the plan of a room to be arched, *e* is a half circle, or the given arch, which *f* and *g* are taken from it; divide the front Bracket *e*, into 8 equal parts, or more, the more parts the better; divide *f* into the same number of parts; then take 1—2, 3—4, 5—6, 7—8, from *e*, and set them on to *f*, 1—2, 3—4, 5—6, 7—8; then tack in nails at 1, 3, 5, 7, &c. and bend a thin slip of a board round those nails, and mark round by it, and this Bracket will exactly mitre with the Bracket *e*. *g*, is an Angle Bracket; draw the Base line, and divide it into the same number of parts as the given Bracket *e*; then take 1—2, 3—4, 5—6, 7—8, from *e*, and transfer them to 1—2, 3—4, 5—6, 7—8, on the Angle Bracket; then tack in nails at 2, 4, 6, 8, and bend a slip of a board and mark round as above.

Pl. 24.

Fig. 1.

Fig. 2.

Fig. 3.

Pl. 25.

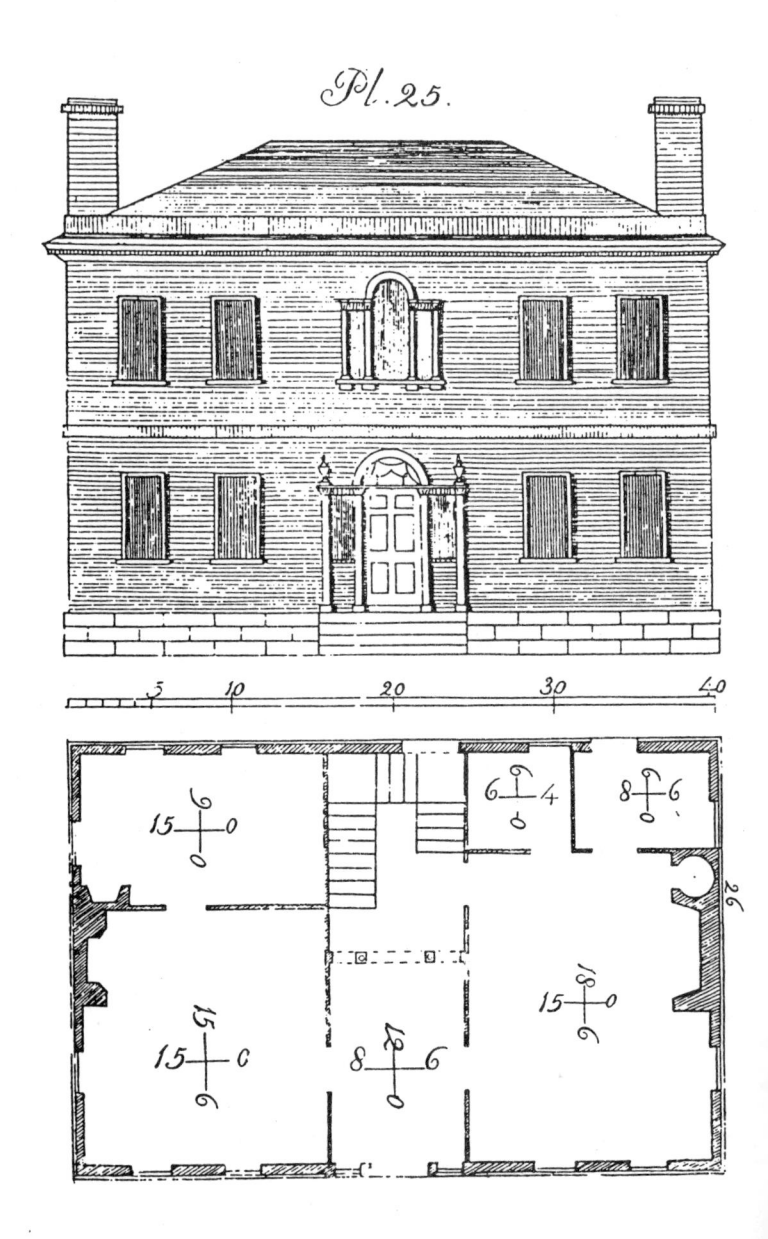

Pl. 26.

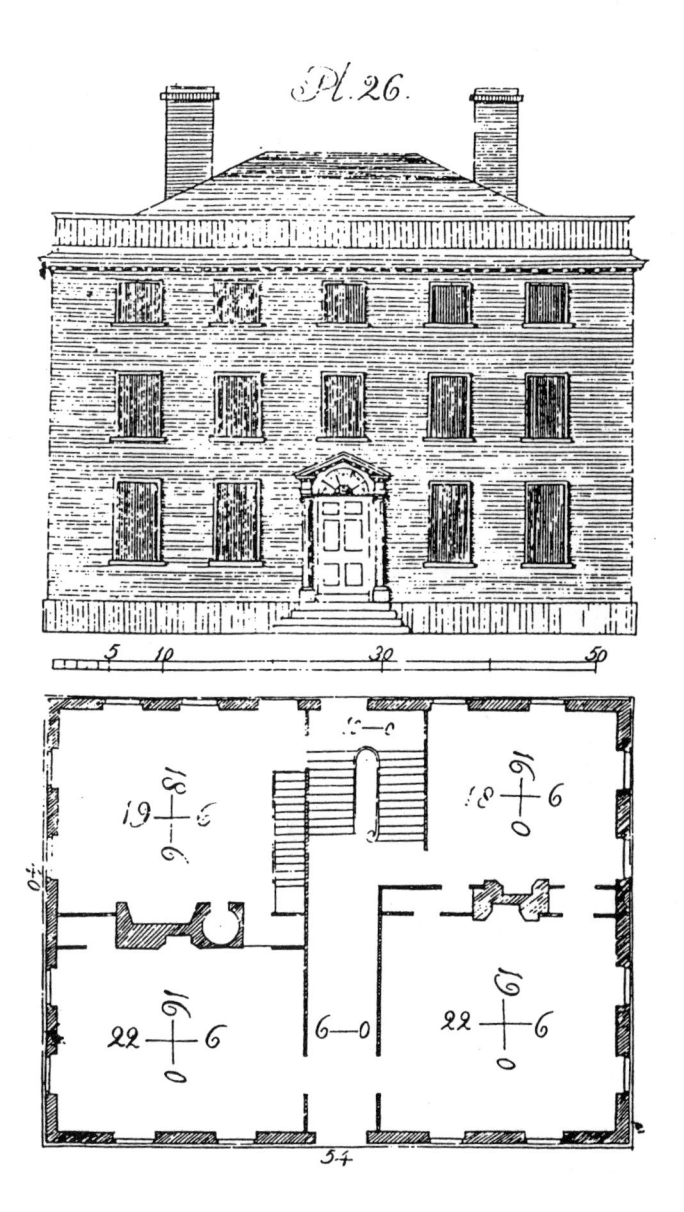

Plan of a Pulpit

Pl. 28.

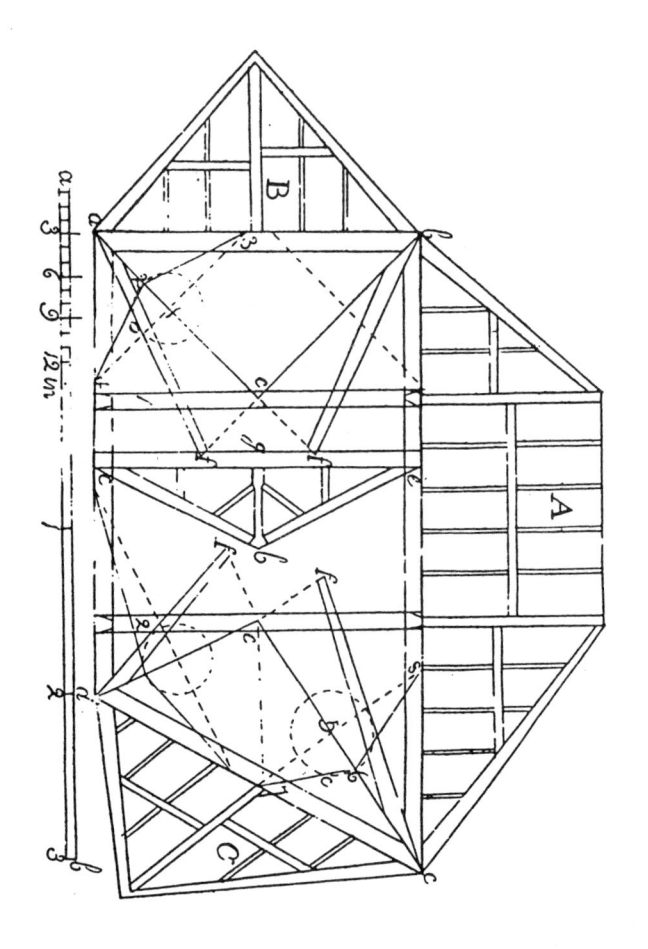

PLATE XXVIII.

The Plan of a Roof in Ledgment, showing the method to find the length of the Hips, Square or Bevel, and their backing to any pitch required.

LET *a*, *b*, *c*, *d*, be the angles, or corners of the Building, to find the length of the Hips, and their backing: first, Lay down the plan of the Roof, *a*, *b*, *c*, *d*, to a scale of one inch to a foot, as the scale *a*, *b*; then draw the principal Rafters on the plan *c*, *b*, *c*, and dispose of the beams at proper distances, as room will admit, which beams, number 1 and 2, will stand to receive the top of the Hips; then draw the base line of the Hips, *a*—*c*, *b*—*c*, at the square end, and at the bevel end *c*—*e*, *d*—*e*; then take the perpendicular height of the principal Rafters, *g*, *b*, and set it perpendicular, from the base line of the Hips, *a*—*c*, *b*—*c*, and *c*—*e*, *d*—*e*, as *c*, *f*, and *e*, *f*; then draw the lines *a f*, *b f*, and *c f*, *d f*; these lines will be the length of each Hip respectively; then to find the backing of the Hips, draw a line square with the base line of the Hips, as 3, 1, 4, and 7, 5, 8; then set the compasses at 1, and extend them to touch the Hip at *o*, and draw the small dotted circle, as there described; then from the point 2, where that circle cuts the base line, draw these lines 2—3, 2—4, which are the backings of the Hips; proceed in the same manner at *b*, *c*, and *d*, as will appear plain to every practitioner, on inspection.

B, A, and C, are ends and side in Ledgments.

PLATE XXIX.

FIGURE A, is a front Bracket. Fig. B, is an Angle Bracket; transfer from A to B, as the figures direct, from 1—1, to 1—1; 2—2, to 2—2, &c. C, is a roof; divide the width of the building into 4 parts, one of which will be the perpendicular height. Divide Fig. D, into 7 parts, give two to the perpendicular height.

Fig. E, is intended for a roof to a Meetinghouse; divide the width of the building into 9 parts; give two to the perpendicular height; the ends of the Beams, *a, a,* are to be supported by columns.

Pl. 29.

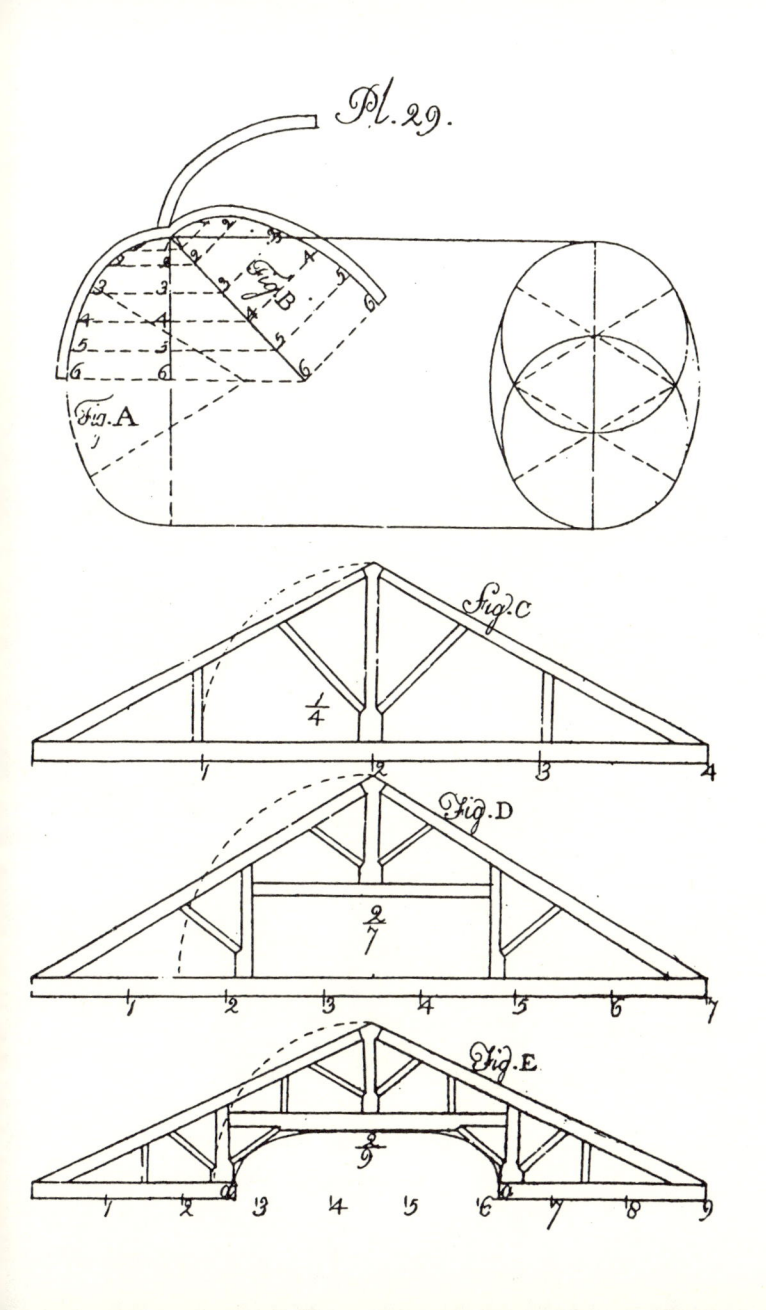

Fig. A
Fig. B

Fig. C
1/4

Fig. D
2/7

Fig. E
2/9

Pl. 30.

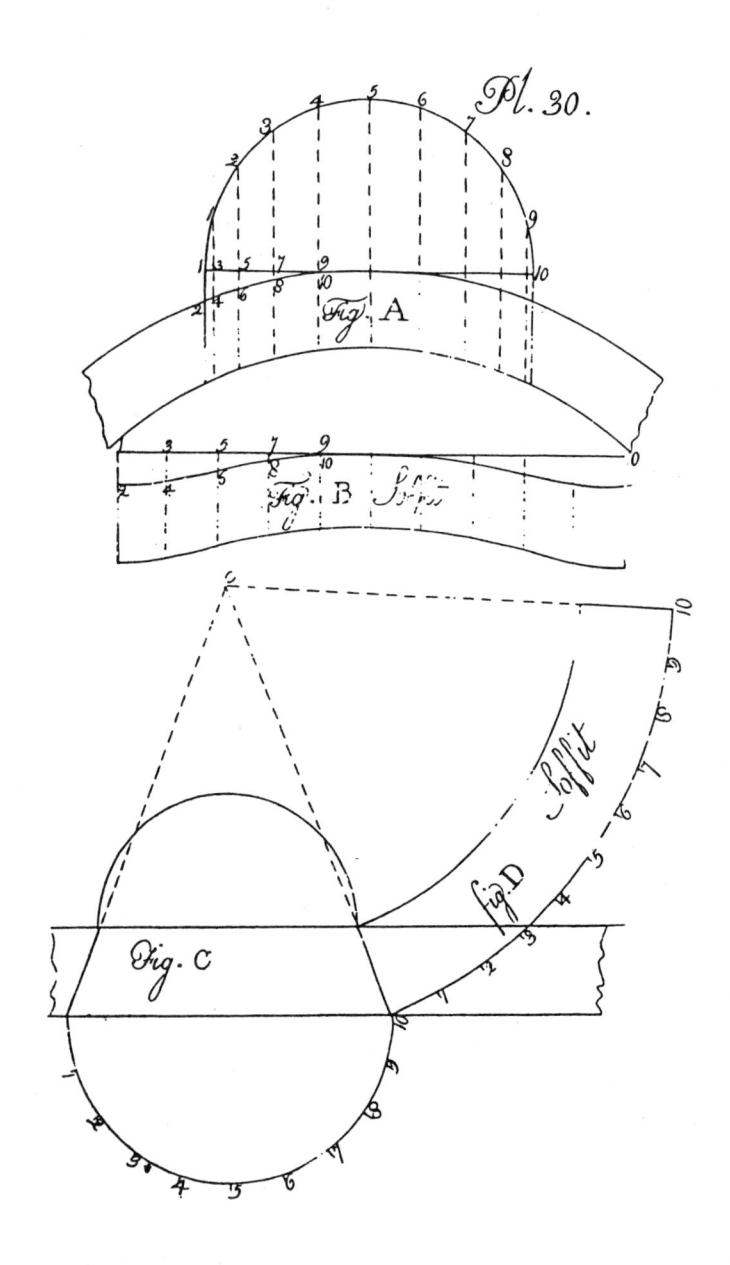

Fig. A

Fig. B

Fig. C

Fig. D

PLATE XXX.

FIGURE A, the plan of a circular Wall, wherein a circular door or window is to be fixed. To make a soffit to fit or stand on the plan as Fig. B, draw the base line of the arch or soffit, to touch the bow of the wall; divide the arch line into ten parts, and drop them down to the plan acrofs it; then ſtretch cut the arch, as 1, 10, and draw the divifions at right angles from it; then take them from the bafe line to the wall, as 1—2, 3—4, 5—6, &c. and transfer them on the parts of the line ſtretched out at B, 1—2, 3—4, 5—6, &c. that will give you the edge of the foffit B.

FIG. C, is a circular foffit in a ſtraight wall, on fluing jams; draw the fluing of the jams until the lines interfect as at *e*, that will be the centre for drawing the edge of the foffit, which is plain to infpection.

APPLEWOOD BOOKS
BRINGING THE PAST ALIVE

*TIMELESS ADVICE & ENTERTAINMENT
FROM AMERICANS WHO CAME BEFORE US*

George Washington on Manners
Benjamin Franklin on Money
Lydia Maria Child on Raising Children
Henry David Thoreau on Walking

&

Many More Distinctive Classics
Now Available Again

At finer bookstores
& gift shops or from:

APPLEWOOD BOOKS
18 North Road
Bedford, MA 01730